Look and Learn

by Layne deMarin

Table of Contents

Consultant:
Adria F. Klein, Ph.D.
California State University, San Bernardino

capstone
classroom

Heinemann Raintree • Red Brick Learning
division of Capstone

Take a Good Look

Scientists learn a lot just by watching the world. They notice how animals move, how the weather changes, and what happens during a flood.

This scientist is watching birds. She can watch them without scaring them by using **binoculars**. Scientists use other special tools that help them see things up close.

In Space

Binoculars are not strong enough to help scientists see what is in outer space. Stars and planets are thousands, even millions, of miles away from Earth. Scientists have to use **telescopes** to see into space.

This is the moon. This is what it looks like when you look at it through a telescope. You can see more details looking at it this way.

This is what a new star looks like as it is forming. This star is very far away from Earth. Scientists used a space telescope to take this picture.

Scientists send special telescopes and cameras into space on machines like this one. The telescopes and cameras send pictures back to the scientists on Earth. Then the scientists can see what things look like in outer space.

At Small Things

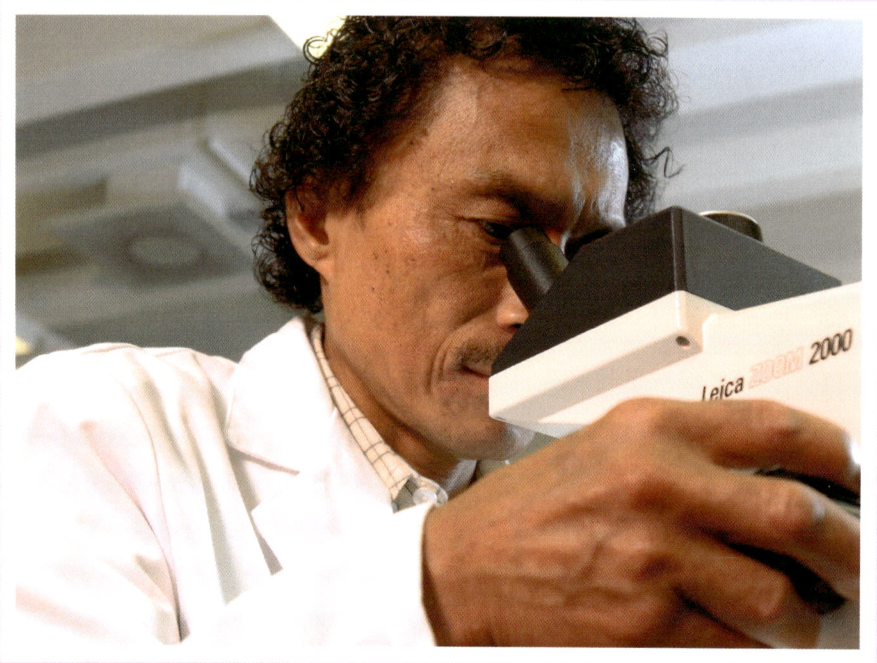

Some scientists study the smallest things in the world. Binoculars and telescopes won't work for that. Scientists use a **microscope** to make tiny things look bigger.

Using a microscope lets scientists see a butterfly wing up close. That is how they can see that the wing isn't smooth but covered in scales.

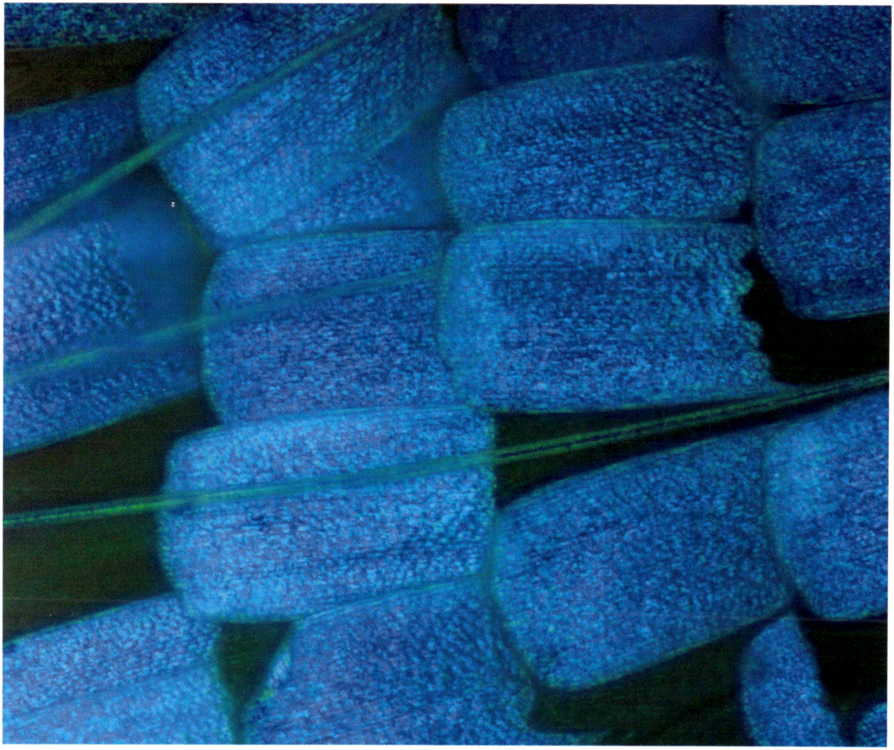

Look at this pond. Your eyes can tell you some things that are true about this pond. Scientists will use a microscope to find out more about the water in the pond.

These are **algae**. They live in the water in the pond. You can see them by looking through a microscope. Millions of algae live in just one drop of pond water. They make the water in the pond look green.

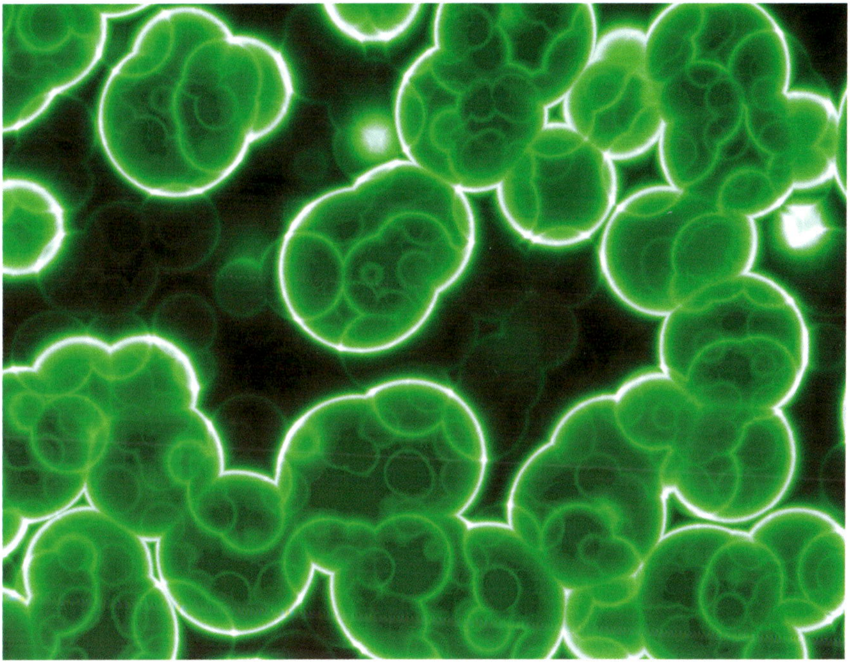

Inside

Doctors are scientists who learn how our bodies work. Doctors help people stay healthy by understanding what is happening inside their bodies. One way to do that is to use an **X ray**.

Another tool doctors use is an **MRI machine**. This shows them a different view of the inside of a person's body.

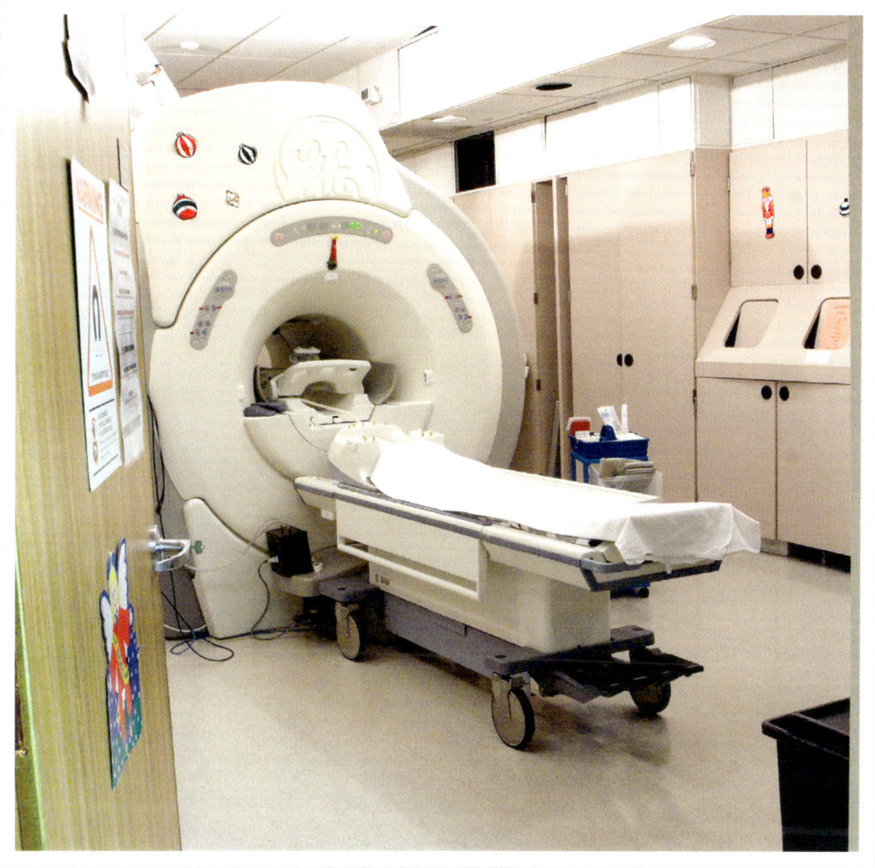

At the World

You can think and explore like a scientist does. You can look at the world carefully to learn more about how it works. Just like a scientist does, you can learn things by using a computer.

You can do science experiments. You can watch what happens. Then you can share what you have learned. You can look and learn!

Glossary

algae small plants without roots or stems that grow in water

binoculars a tool that you look through with both eyes to make things far away seem closer

microscope a tool that you look through with one eye to make small things seem bigger

MRI machine a tool that lets doctors see inside the human body

telescope a tool that makes things far away seem larger and closer

X ray an invisible beam of light that is used to take pictures of things inside the body